A very special thank you to Logan Burkett for his incredible editing skills and to all my friends and family who supported me.

Thank you and never stop pursuing your dreams!

Rachel Lassman

To Alyssa and Brianna!
Believe in yourself always!

"Teacher!" said Billy.
"Look what I found.
It's red with black spots.
And its body is round."

"A ladybug, Billy!"
Said his teacher with glee.
"What a wonderful find.
Can you bring him to me?"

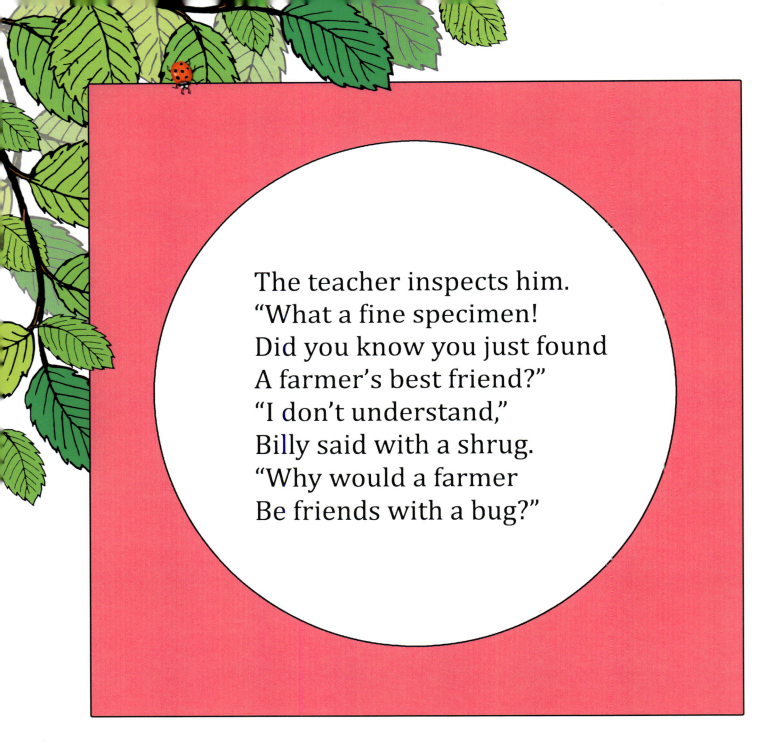

The teacher inspects him.
"What a fine specimen!
Did you know you just found
A farmer's best friend?"
"I don't understand,"
Billy said with a shrug.
"Why would a farmer
Be friends with a bug?"

"Well, ladybugs help them
Get rid of some pests.
Come sit down beside me
And I'll tell you the rest.
First, you must know that,
Despite what they're called,
They're actually beetles.
They're not bugs at all!"

"A bug and a beetle
Do different things.
They have different mouths
And they have different wings.
The mouth of a true bug
Is skinny and long.
It's used to suck liquids
And acts like a straw."

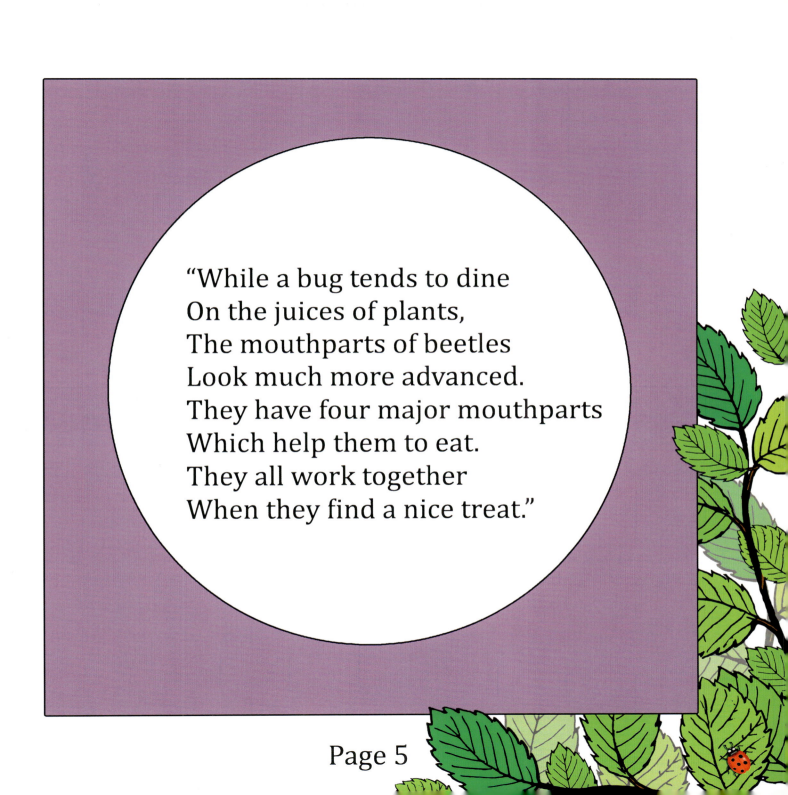

"While a bug tends to dine
On the juices of plants,
The mouthparts of beetles
Look much more advanced.
They have four major mouthparts
Which help them to eat.
They all work together
When they find a nice treat."

"The mandibles open
And close side to side.
They act like their teeth
As they chew, pinch, and grind.
They have modified lips,
Both on bottom and top,
Which helps hold in their food
As they constantly chomp."

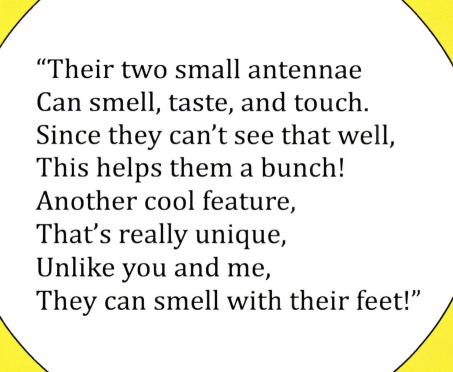

"Their two small antennae
Can smell, taste, and touch.
Since they can't see that well,
This helps them a bunch!
Another cool feature,
That's really unique,
Unlike you and me,
They can smell with their feet!"

"They're also quite tidy,
And after they're fed,
They use their front legs
To clean off their head."

"That's amazing!" said Billy.
"The differences between,
All the beetles and bugs
That I've never seen."

SOAPY
BUCKET

"That's right," said his teacher.
"But, there are differences in
All ladybug species,
So let me begin.
Over 5,000 species
Call our Earth home.
There are 400 species
In the U.S. alone."

"They all can look different.
They don't have to be red.
Some are orange, pink, or blue,
Some are brownish instead.
Not all of the ladybugs
Will always have spots.
Some will have many
While others will not."

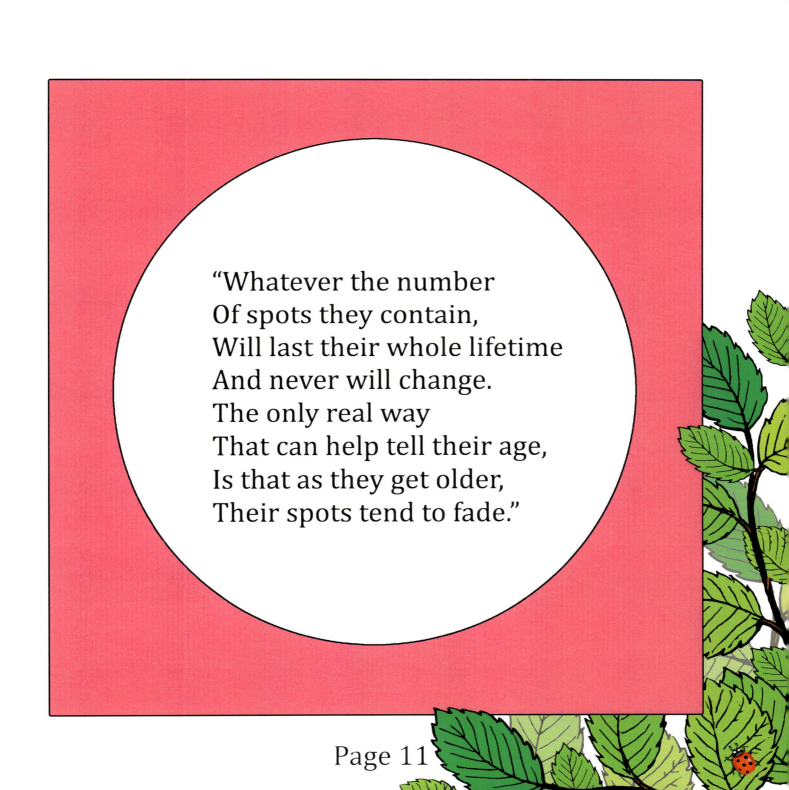

"Whatever the number
Of spots they contain,
Will last their whole lifetime
And never will change.
The only real way
That can help tell their age,
Is that as they get older,
Their spots tend to fade."

"Another neat thing,
For which most aren't aware,
Is that ladybugs have
Similarities to bears!"

"Bears?" hollers Billy.
"Please, tell me how?
The two are so different.
I need to know now!"

How am I like you?

"Well, they both dislike cold
And when temperatures drop,
They start to look for
A really warm spot.
Once they have found it,
Their perfect warm place,
They'll sleep through the winter.
This is called hibernate."

"For about three whole months,
The ladybugs sleep.
They never wake up
To play, drink, or eat.
They can live off their fat
That their bodies have stored,
And sleep close together
To stay nice and warm."

"That's amazing to know!
But, you still haven't said,
What makes this cool beetle
A farmer's best friend?"
"That's a great question, Billy,
And I must conclude,
That the reason they're friends
Is based on their food."

"Ladybugs dine on
A tiny green bug,
Which feeds on a plant's
Young leaves and buds.
This bug's called an Aphid,
And they do so much harm
To all kinds of plants
That are found on a farm."

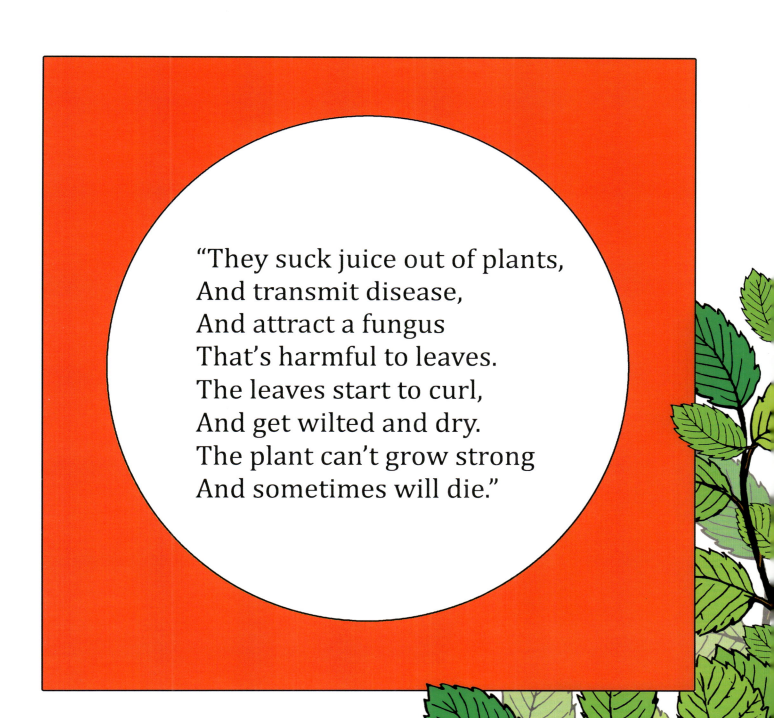

"They suck juice out of plants,
And transmit disease,
And attract a fungus
That's harmful to leaves.
The leaves start to curl,
And get wilted and dry.
The plant can't grow strong
And sometimes will die."

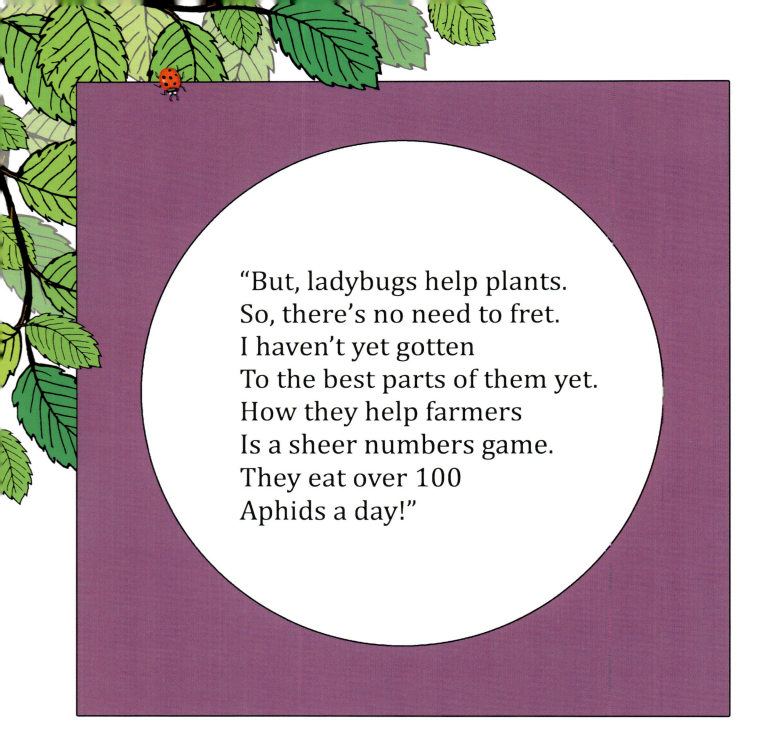

"But, ladybugs help plants.
So, there's no need to fret.
I haven't yet gotten
To the best parts of them yet.
How they help farmers
Is a sheer numbers game.
They eat over 100
Aphids a day!"

"Just a few ladybugs,
A small group of nine,
Could eat over a million
In just one lifetime.
They can lessen the need
For bad pesticides
And help you and me
Live healthier lives."

BUCKET
-O-
APHIDS

"That's amazing!" said Billy.
"How they help us so much.
Simply by eating
A large portioned lunch!
There's still one more thing
That I don't understand.
Most insects I've seen
Blend in with the plants."

"How does a ladybug,
Being so red and bright,
Stay hidden and safe
From a hungry bird's sight?"
"I'm glad that you asked that.
There are three things they do
To keep themselves safe.
Let me tell them to you."

"Bright colors in nature
Warn predators, 'Stay Clear.'
You do not want to eat
The snack you have here.
They might have a poison,
Or a horrible taste.
It's best to keep moving,
And just stay away."

"A ladybug can also
Pretend to play dead,
Since predators prefer
To eat live prey instead.
If the acting and colors
Don't stop the attack,
There's one final defense
From becoming a snack."

WARNING!!!!!
LADYBUG BLOOD

Terrible Taste

Terrible Smell

Makes You Sick

"A ladybug oozes
Its blood from its legs.
A yellowish substance
That works in three ways.
Its taste is quite awful.
Its smell even worse.
Making most predators
Want to run in reverse!"

"But, if for some reason
They still want a lick,
Their blood has a poison
That makes predators sick.
They'll remember this sickness
For the rest of their lives
And run for the hills
When a ladybug comes by."

"I can't believe there's so much
That ladybugs do.
There are so many facts
That I never knew.
They help out the farmers
With the aphids they eat.
They are actually beetles
And they smell with their feet!"

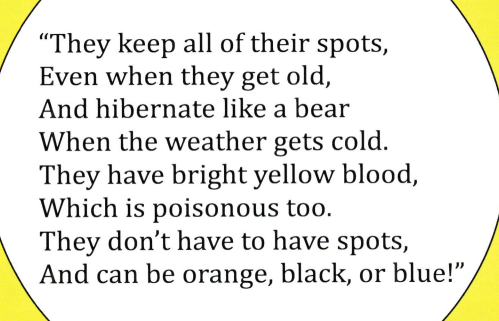

"They keep all of their spots,
Even when they get old,
And hibernate like a bear
When the weather gets cold.
They have bright yellow blood,
Which is poisonous too.
They don't have to have spots,
And can be orange, black, or blue!"

"There are so many things
That make ladybugs great.
From the cute way they look,
To the rest of their traits.
I'm so happy I found one,
And brought him to you,
So you could teach me about
All the things that they do!"

Favorite Ladybug Facts: